HOW PACKAGING CAN IMPROVE MANUFACTURING OPERATIONS

Sterling Anthony, Jr.

AMA Management Briefing

AMA MEMBERSHIP PUBLICATIONS DIVISION
AMERICAN MANAGEMENT ASSOCIATIONS

Library of Congress Cataloging in Publication Data

Anthony, Sterling, 1949-
How packaging can improve manufacturing operations.

(AMA management briefing)
1. Packaging. 2. Materials handling 3. Production management. I. Title. II. Series.
TS195.2.A58 1983 658.7'88 83-21547
ISBN 0-8144-2298-5

This Management Briefing has been distributed to all members enrolled in the Packaging Division of the American Management Associations. Copies may be purchased at the following single-copy rates: AMA members, $7.50. Non-members, $10.00. Students, $3.75 (upon presentation of a college/university identification card at an AMA bookstore). Faculty members may purchase 25 or more copies for classroom use at the student discount rate (order on college letterhead).

First Printing

About the Author

Sterling Anthony, Jr., is president of Sterling Anthony, Inc., a consulting firm based in Detroit. He specializes in the marketing, production, and distribution aspects of the packaging function. He also serves as program director of the Institute for Packaging Advancement, a nonprofit firm that conducts research and sponsors educational seminars on packaging. Mr. Anthony was formerly employed with Kraftco, Abbott Laboratories, Johnson & Johnson, and Ford Motor Company.

His professional credentials include certification by the Society of Packaging and Handling Engineers, the International Material Management Society, the National Association of Purchasing Management, the American Production and Inventory Control Society, and the American Society of Traffic and Transportation.

Mr. Anthony received a B.S. degree in packaging from Michigan State University, where he is currently a faculty member in the school of packaging. He holds an M.B.A. in marketing and finance from Roosevelt University and is now a Ph.D. candidate in marketing at Michigan State.

Contents

Introduction

Packaging is a complex and versatile function. It cuts across numerous functional lines and, as such, influences the multiple aspects of company dynamics. Its relationship with manufacturing and the manufacturing mission—briefly, to produce products that are effectively made and effectively packaged—places the packaging function in a position to have substantial impact on a company's profitability. More broadly, packaging can be viewed as one key to improving the nation's production system as a whole.

A grandiose claim for a humble function? Not when we look at the packaging/manufacturing relationship in the framework of national and worldwide trends.

THE SCENARIO: AN ECONOMY IN TRANSITION

Indeed, it is now apparent that the economy of the United States is embarked on a chartable transition from a manufacturing base to a service base. This passage is sometimes taken to signify the decline of manufacturing's importance to the growth and stability of the U.S. economy. On the contrary, what needs to be understood is that the transition parallels the shift from an agricultural to a manufacturing economy that occurred earlier in this century. As the number of people who made a living at farming decreased, the number of people entering the manufacturing industries increased. Just as technology permitted fewer farmers to feed the masses, technology will permit fewer industrial laborers to produce the tangibles our society demands. And, just as food became no less important merely because fewer people were producing it, manufacturing and the quality of life it provides will not decline in importance simply because of decreasing labor ranks. We still need to eat. And we still consume products, and will continue to do so.

The manufacturing segment will continue to contribute a large portion of our economic wealth. Instead of its role diminishing, manufacturing will become an increasingly vital link in our attempt to solve broader economic problems.

The problems are boldly and unequivocally manifest:

- Low productivity
- High cost of resources
- Worldwide competition in a global marketplace.

With nothing less than our survival as a world-class nation at stake, the manufacturing mission is more important than ever. At present, reformulation of that mission is essential. It must be done in a manner that confronts and seeks to counter the disturbing global conditions that threaten the ability of the United States to compete internationally.

"Reindustrialization" is one of the terms coined to describe the plan by which the manufacturing/industrial sector will take on the responsibilities and challenges posed by current and future circumstances. To the layman, the term means a restructuring of our industrial capabilities so that we can manufacture with greater productivity, greater resource efficiency, and greater quality, the results being greater sales, profits, and market share.

A cornerstone or reindustrialization is the systematic integration of manufacturing with other key functions of business, such as design, engineering, marketing, and finance. That effort at integration, usually attempted with a computer, has given birth to such concepts as CIM (Computer Integrated Manufacturing), CAD/CAM (Computer Aided Design/Computer Aided Manufacturing), the Automated Factory, and Flexible Manufacturing.

Manufacturing now occupies—or indeed should occupy—the center-stage position in the minds of enlightened management. The need for reindustrialization—and the systematic integration on which it depends—point, in turn, to the need to identify and improve all major manufacturing interfaces.

THE PACKAGING PERSPECTIVE

One such interface, the subject of this briefing, is that between manufacturing and packaging. Since packaging intersects many functional lines, packaging decisions often represent compromises among numerous and varied departmental considerations. Such compromises should be achieved objectively and systematically. They should reflect the concerns of each function, but with a clear sense of which concerns claim greater priority for the overall betterment of the company. Manufacturing-related requirements, quite justifiably, will often take top priority. And it is important that these concerns be accorded a fair level of influence on packaging decisions.

Typically, a packaging operation is seen as something that happens at the end of the manufacturing line. The former receives the output of the latter. In fact, however, packaging impacts the entire system. In other words, the interface between the two is not limited to in-plant packaging operations. It should extend to the packaging of incoming shipments and the containerized movement of goods through the plant. And, of course, packaging must protect the product through the distribution process.

This briefing is devoted to discussions on ways in which to use packaging to improve manufacturing operations. It is based on the premise that manufacturing and packaging are both components of the sourcing-production-distribution continuum that is industry itself.

A company's objective is to produce products that are effectively manufactured and effectively packaged. Its ability to do so is in direct accord with how well the manufacturing and packaging operations have been integrated.

1

A Packaging Primer

What is packaging? The question can evoke all kinds of answers, depending on whom you ask. Since packaging is a multifaceted discipline, it is often hard for individuals with different responsibilities to obtain a bird's-eye view of the field. This chapter presents such an overview: a perspective from which to understand the packaging function and its relationship to the whole and to all of its parts, particularly manufacturing.

PACKAGING IS A SYSTEM

The manufacturer who stands to make the most of the packaging contribution is the one who understands the systems approach. That involves extending all decision-making considerations *beyond* the plant to include conditions that precede and follow the manufacturing operation.

Revolutionary ideas, from Galileo's cosmology to Einstein's unified field theory, see things in terms of a system. A system, by definition, is a whole consisting of component parts. The synergism usually inherent in a system means that the integrated whole is greater than each of its parts. Each component part is interrelated with each other part; what is done to one component impacts on all the other components.

Trade-offs among the requirements of the various components are at the heart of systems management. The objective is not to attempt to optimize any particular component, but to manage the trade-offs so that the net result enhances the whole.

The packaging system begins with the retrieval of basic raw materials such as metal ore, sand, wood, or petrochemicals. These materials are converted into formable packaging materials, and then into packaging containers and components. The formed packaging is then combined with the product. The now-packaged product may undergo further packaging, in the form of cartoning, boxing, and unitizing. The packaged product then proceeds through distribution channels, eventually reaching its final destination in the hands of industrial or retail/consumer customers. Disposal and/or recycling of the packaging signals the end of the system cycle.

Joining the manufacturer of the packaged product as parties to the system are raw material producers, converters, machinery builders, distribution channel intermediaries, the customer, and a host of services such as consultants, designers, and disposal firms.

In addition to being a system within itself and a component of a larger company system, packaging is a part of an even greater system: The manufacturing/packaging team is a microcosm of the larger flow of goods within U.S. industry. Sourcing provides the materials from which to manufacture products, the manufacturing industries produce the products, and marketing/distribution segments disseminate them. It is within this larger context that the packaging and manufacturing relationship can be managed for sizable gains.

THE FUNCTIONS OF PACKAGING

Despite the enormous diversity of packaged products, packaging performs only four functions: *containment, protection, communication*, and *utility*. *Containment* refers to packaging as a receptacle, a holder of its contents. *Protection* emphasizes packaging as a barrier between the product and conditions that would damage or degrade it. In the example of hazardous materials, protection extends to shielding people and the environment from what could be harmful effects of the products. *Com-*

munication refers to packaging as a conveyor of required information, usually about the nature of its contents. *Utility* is a broad-based concept. It describes the convenience features provided by the packaging such as easy opening, transportability, machinability, and disposability.

Packaging development is the formal, systematic procedure wherein the developer attempts to incorporate into the package the optimal blend of the four functions, after weighing information gathered to determine the importance and priorities of all factors and all functions.

THE PACKAGING ARENA

The packaging system and how it works cannot be understood without an appreciation of the dynamic nature of the function and the field as a whole. Innovative improvements in packaging materials, machinery, and services are the established rule, rather than the exception. The result is that a firm can find itself at a competitive disadvantage because it is not making optimal use of technological advancements. Granted, no firm can adopt every new change that comes down the pike. Economics, return-on-investment requirements, and other practicalities are constraints on a firm's ability to adopt new offerings in the packaging field. But keeping a watchful eye is important.

The only rational response to this ever-changing face is to commit the firm to a *planned* approach to the management of packaging. The emphasis must be on planning for the future, not simply scanning the short-term horizon. A strategic, long-term view is essential. This is where the systematic and integrative approach mentioned earlier comes in. The approach calls for a *systems* view, which in turn requires systematic, integrative planning. In this planning, top-level strategies direct the course of the firm, manufacturing strategies support top-level plans, and packaging plans support those of manufacturing.

The firm should view its packaging needs within the context of the overall direction established for the enterprise. These are the questions it should be asking:

- What products will we develop over the next five, or even ten years?

- What are our life-cycle plans for existing products?
- What new markets will we penetrate?
- How should we define the competition, present and future?

The answers derived from such penetrating analyses provide the foundation of the stewardship of the firm—its responsibility to manage the concerns of the enterprise with proper regard for the integrity of each of its functions. This kind of inquiry is also highly relevant to effective packaging management. The answers form the framework for strategic and packaging plans, including whether to adopt programmed packaging and how to improve supplier relations, formulate policies, recruit personnel, project production capacities, and so on.

On the basis of this perspective, the next chapter will explore in greater depth the subject of planning, stressing the strategic areas that include both manufacturing and packaging considerations.

2

Planning Manufacturing and Packaging Operations

The fundamental purpose of manufacturing is to receive materials, components, and assemblies, and add value by converting them into finished products. The systems approach means recognizing this flow, or continuum concept. It also means that manufacturing cannot meaningfully fulfill its mission by attempting to isolate its concerns from those of the larger complex of business activities.

As is true of all functions in an organization, manufacturing operations create a "ripple effect." The widening circles of influence eventually touch on the broadest-reaching margins of the company's operations. Thus, the company must accord manufacturing strategy a position at the highest planning level of the firm. Similarly, packaging plans and decisions should support the ultimate manufacturing mission shaped by this strategy.

EXAMPLES OF THE INTERDEPENDENCE OF PLANNING

Consider two common business planning strategies: competing on the basis of price and competing on the basis of product differentiation.

Each carries manufacturing- and packaging-related ramifications, which if misunderstood or ignored will seriously undercut the firm's prospects for success.

The firm that competes on the basis of price-competitive strategy is highly dependent on manufacturing. Production costs (manufacturing plus packaging) must be kept low in order to sell at a low price. Low production costs are mainly achieved by means of automated processes. Automated processes are optimized by long uninterrupted runs, which are in turn best achieved with a minimum of production-line changeovers. These operations place a premium on product standardization. Still another requirement of high-volume automated output is a market share large enough to absorb the output. A price-competitive strategy, therefore, must be supported by the right manufacturing strategy, which will affect such areas as the versatility of product lines, new product development, production flexibility, and market share.

PRODUCT-DIFFERENTIATION STRATEGY

The firm that competes on the basis of product differentiation also relies heavily on manufacturing. Product differentiation can take the form of broad product lines, better quality, faster and more dependable delivery, and so on. Manufacturing sequences used in product differentiation can also vary, depending on many factors. As a rule, however, differentiated products require high inventory levels, higher quality standards, more frequent production-line changeovers, and greater emphasis on new product development than the nondifferentiated, standardized products that compete mainly on the basis of price. Differentiated products are more expensive to manufacture. How much more expensive is, to a large extent, under the control of the manufacturer.

HOW IS PACKAGING AFFECTED?

Regardless of the manufacturing strategy applied, certain production concerns that relate to packaging should occupy management's thoughts. Basically, they tend to revolve around the costs of forming, filling, and closing the packaging on the manufacturing floor. There—on the floor—each decision carries a consequence.

Take the price-competitive product. Although the truism that a company must produce at a low cost in order to sell at a low price cannot really be questioned, the related issue of "how much economy" the firm can afford in its packaging also has to be addressed. Remember that the low-price product will most likely be manufactured and packaged on automated equipment. Low-cost packaging that is of matching low quality can cause manufacturing problems resulting from breaking, jamming, equipment failures, and the like. The consequence is expensive downtime.

The differentiated product often uses packaging as the basis of its differentiation. In subscribing to the "integrated product" concept, the planning team will tend to view the total product as composed of the actual product *and* its packaging. The decision-makers, then, have *two* variables to consider in coming up with a new and differentiated product.

With differentiated products, we have the setting for one of the classic conflicts involving packaging: that between manufacturing and marketing. In its verve for differentiated products, the marketing department sometimes imposes requirements on packaging that run counter to the desires of the manufacturing department. The shape and construction of the packaging so cherished by marketing can be loathed by manufacturing if it drives up production costs through slower running speeds, more labor requirements, more frequent quality checks, and the like. Conflicts are inevitable. The effectiveness with which they are resolved depends on how systematically the problem-solver approaches the problem.

PLANNING—A LEAP OF FAITH

George Odiorne, in his *Management by Objectives* (1965), pointed up the loss of entrepreneurial values, specifically the willingness to take calculated, intelligent risks, that has occurred in the business community over the course of the past decades. Bureaucracy and glorification of the annual report, he asserted, had drained the vitality of individual corporate leaders and discouraged the ingenuity and boldness that once propelled American business.

Now, perhaps more than at any time since the Depression, the pressures of the economy have businesses in a strangulating double bind that makes Odiorne's analysis even more relevant today. At a time when industry needs to invest in productivity-increasing and cost-reducing projects, finances constitute a formidable obstacle. High interest rates and tight money have been the bane of industry in the recent past. Scarce and costly capital requires a higher return on investment to justify its deployment. The timing could hardly be more unfortunate.

Business seems to be divided into two camps. One is populated by companies in the grip of decision-making paralysis, who have tabled all considerations for improvements in facilities, processes, and equipment until the return of the mythical "better days."

The other camp is home to companies that are pledged to deal with current conditions. They perform penetrating analyses of their companies and environments to unearth investment opportunities, subjecting these options to tough screening criteria and realistic financial analysis. The projects so selected are implemented according to a thought-out management plan. The economic worth of plans and proposals—professionally quantified, detailed, and presented—can then meet the sharp scrutiny that any top management must necessarily apply.

Packaging decisions and long-range plans call for the same kind of tough-minded scrutiny. A number of current conditions will form the parameters of packaging plans for investment.

Productivity. Disturbingly down from past levels, productivity is a national concern. At stake is the ability of the United States to compete successfully in world markets and, without exaggeration, to survive economically.

Volumes upon volumes have been written in an attempt to identify the causes and cures of our productivity ailments. Without trying to critique and summarize that body of work, this assertion can be made: If productivity—the ratio of outputs to inputs—is to improve appreciably, an essential area of concentration has to be the manner in which industry manufactures its products. Packaging operations are part of manufacturing operations. Logically, then, plans to upgrade productivity through better manufacturing should give due consideration to the role of packaging.

Automation. Closely allied to the issue of productivity is that of automation. The reader is undoubtedly familiar with the often-cited benefits of automation: lower production costs, higher productivity, better control, and improved quality. The reader may have given less thought to the fact that automated manufacturing operations impact packaging operations: More sophisticated production machinery calls for more sophisticated packaging operations. Incompatibility of the two can result in numerous bottlenecks in the manufacturing process.

Energy. These are still energy-conscious times. Energy-related factors influence decisions on manufacturing locations, facilities, equipment, and so forth. In the same way, packaging is affected by energy. Packaging materials and containers, for instance, differ in terms of the energy required to manufacture, transport, fill, and dispose of them. Similarly, energy-consuming differences distinguish packaging equipment. (Shrink wrap vs. stretch wrap is a good example.) Energy is recognized as a long-term strategic element in a firm's future. Energy-efficient manufacturing and packaging systems pay dividends on a daily basis.

Shortages. Finite resources are an external constraint, but, until recently, the industrialized Western world did not have to seriously contend with the consequences of that fact. Today, shortages—artificially created or not—are a concern that industry is forced to deal with, like it or not.

In choosing a packaging material, for instance, management must make the future availability of the material and its susceptibility to shortages and supply interruptions a major consideration. This concern impacts on the role of packaging in the coming automated era. It would prove highly expensive for equipment to be down because packaging material were unavailable or because the material incorporated poor production performance characteristics. The latter point underscores the importance of integrating the choice of material with the choice of equipment and systems through which the material will run. Also, since automated systems perform best with standardized components that call for tightly held tolerances, changeovers from a specified material may not be a simple matter. Conceivably, changes in the production system might be necessary to accommodate the new material. Foresight in choosing materials is essential.

Competition. Competition has become increasingly international. Domestic manufacturers are being compelled to meet the challenge of foreign competition at home. Additionally, a product carrying the label of a U.S. manufacturer may well have been made domestically from components sourced overseas, manufactured overseas, or distributed overseas.

The global condition means that better manufacturing and packaging operations that result in lower costs and better quality will enable companies to compete more strongly in international markets. For the international firm, standardization of operations will serve to minimize the complicating effects of doing business across national boundaries. The protective integrity of the packaging is another supremely important aspect. International shipment of packaged goods represent a hazardous journey. A shipment of damaged goods subtracts from customer goodwill and the firm's attempts to penetrate foreign markets.

Product development. Today's customer is more sophisticated and discerning than his predecessors were. To compete successfully for customer buying power, products must offer benefits that are perceived to be substantial. That carries weighty overtones for product development. The truly innovative product is an infrequent occurrence. Most so-called new products are merely different twists of an existing theme. That probably explains, in part, the low success rates of new products.

High risks notwithstanding, industry must still stake a certain portion of its future on new product development. But in recognition of those risks, industry is placing an increased emphasis on extending the profitable lives of existing products. A direct outcome of that approach will be a higher manufacturing consciousness. Manufacturing and packaging efficiencies that generate cost savings, increased quality, and greater productivity will be sought to enable product winners to remain so longer.

3

Packaging Standardization and Manufacturing

The previous chapter described manufacturing in terms of a flow concept: materials move into, through, and out of a manufacturing facility. During that journey, materials convert from raw materials, to in-process goods, to finished products. Traditionally, a firm's packaging management efforts are concentrated around the finished products.

There are opportunities, however, to improve manufacturing operations through the better management of the packaging of the raw materials and in-process goods. Those opportunities are best exploited when the firm understands the importance and consequences of standardization and develops a coordinated standardization program.

UNDERSTANDING STANDARDIZATION

Standardization is the formal reduction of diversity among items in order to receive quantifiable benefits resulting from volume and/or economies of scale. Applied to packaging, this means reducing the number of different types of packaging in order to make the packaging itself more economical *and* to permit greater standardization of interfacing activities. The interfacing activities are transportation, material handling, and storage. *Transportation* is the activity relied upon to get material to the manufacturing facility and out of the manufacturing facility. *Material handling* is the activity responsible for the in-plant movement of materials. *Storage* is the activity called upon during the planned halts of the flows.

True, standardization permits greater purchase price discounts. But more than that, the more standardized the packaging, the greater the possibility for standardized methods of handling, storing, and transporting the packaging. Standardization of packaging allows less investment in different types of equipment, greater utilization of equipment and space, easier training of personnel, and fewer facility revisions.

But to appreciate the full benefits of this approach, take a closer look at how packaging standardization affects overall operations at each stage of the manufacturing flow. Let's start with the materials that come into the plant.

STANDARDIZING SUPPLIER PACKAGING

Benefits accrue to the manufacturing firm that takes a critical look at the ways its suppliers are packaging the goods shipped to the company. The typical manufactured product is largely composed of materials, parts, and assemblies that are procured from external supply sources. Most supplier shipments are packaged. In fact, supplier-packaged goods commonly constitute most of the volume handled at the receiving dock. This is where materials begin their flow through the manufacturing plant. And the cost and efficiency of the flow is greatly influenced by the supplier's packaging methods. Therefore, it should be the object of management's control.

The best way to attain control of supplier packaging is for the receiving company to establish formal guidelines for the types of packaging that the suppliers may use. The guidelines should be combined into a pamphlet or booklet and distributed to all present and potential suppliers. The actual contents and format of the guidelines would vary among companies and industries. In spite of these variations, every set of guidelines should address a number of important issues. They should, for example, provide the vendor with an explanation of the company's facility and material handling system, preferred channels of communication, and package development considerations. Moreover, the guidelines can specify recommended unit load sizes, and modes of transportation, testing, and labeling.

The objective of the guidelines is to ensure that supplier shipments are packaged economically while providing the necessary level of protection, performance, and compatibility with the receiving facility systems and methods. When such standards are successful, packaged goods will arrive in satisfactory condition. Those same packaged goods will permit quick unloading of the transportation vehicle, easy inspection and verification of shipments, efficient movement to storage or directly to production, and efficient use of space throughout the facility. Each of those benefits can be translated into dollars.

Might such controls on supplier packaging result in higher prices—inasmuch as the supplier may have to depart from his normal practices? This need not be the case. Besides, even if it does happen, it's all right—as long as the price increase is less than the penalty associated with having to receive and handle incompatible packaging. Remember, we must remain committed to a systems approach.

Also keep in mind that the nature of the guidelines are meant to be directional, rather than dictatorial. You want to let the supplier know the type of packaging you prefer, while at the same time, being careful not to totally usurp his responsibility for delivering a satisfactory product to your dock.

It is important to design flexibility into the guidelines. Extremely rigid guidelines are a disservice, in that they obstruct the supplier from exploring more efficient, cost-effective, and innovative ways in which to package the goods. Obstruction can be a considerably costly proposition. Packaging is a dynamic field in terms of its constant parade of new materials, containers, methods, and the like. Recognizing that fact, some manufacturing firms develop a form to be filled out by the supplier. Based on the standard guidelines developed by the manufacturer, this form allows the supplier to provide details related to his proposed method of packaging and shipping.

The geographic dispersion of suppliers is widening, and many shipments are received from foreign locations. This also argues more for standardization, so that the effects of different measurement standards (metric vs. inch), languages, literacy levels, cultures, and the like are minimized.

Regardless of the location of the supplier, unwelcomed conditions can develop: the need to manually transfer nonunitized loads from a vehicle, for example, or to cope with poor identification of goods, or to move loads that are incompatible with receiving dock material handling equipment, or to handle loads that are dangerous due to instability and/or poor quality, or to work with loads that stack poorly or fit poorly into storage racks. All of these conditions represent costly drains on labor, equipment, and space. They are, for the most part, avoidable through better standardized supplier packaging.

TRENDS THAT SUPPORT STANDARDIZATION OF SUPPLIER PACKAGING

The reputation of the U.S. for quality manufactured products has been attacked in recent years. One of the challenges facing American industry is to repair its tarnished image and reinstate itself as the recognized leader in the manufacture of quality products. It stands to reason that the quality of any manufactured product is only as good as that of the materials that go into it. Companies are demanding defect-free deliveries from suppliers. The protective properties of the packaging greatly determine the suppliers' ability to comply. When quality, undamaged materials are received and fed into the manufacturing process, chances are better that a quality finished product will emerge.

Also, consider the trend toward leaner inventory levels. Inventories are costly assets to carry, and companies are searching for ways to trim them. However, when a company ties up less investment in inventory, there is less of a buffer to absorb the effects of damaged shipments. Under such conditions, it becomes imperative that the packaging allow for inventory to arrive in satisfactory and usable condition. I have personally witnessed situations wherein the combination of a lean inventory and the receipt of a damaged shipment brought the manufacturing process to a halt until an emergency shipment arrived.

Yet another trend is automation/computerization: the move toward the integration and unification of major business functions through automation and computer-controlled systems. Receipt of orders, purchasing, design, engineering, manufacturing, quality control, and shipping are all being automated and computer driven. Bar code scanning technology, intimately associated with packaging, is an outgrowth of that trend. Other outgrowths such as automated storage and retrieval systems, automated guided vehicle systems, and high speed conveyorized transport work best with packaged loads that are standardized to the practical maximum.

Lastly, consider the perennial trend toward cost reduction. It's well accepted that a large percentage of the sales price of any product is due to the cost of supplied materials. A company can lower the costs of such items by making sure that the packaging used by the supplier (and charged to the company) is the most economical for the intended performance.

The point, then, is clear: The manufacturing sector benefits by controlling supplier packaging through standardizing guidelines. The materials that are received, handled, stored, and fed into the process have been done so economically, efficiently, and with the minimum sacrifice of quality.

PACKAGING STANDARDIZATION AND IN-PLANT HANDLING

Once received at the docks, packaged supplies have to move to the manufacturing floor—the province of the material handling function. As with receiving operations, packaging will affect how well in-plant handling is performed. Once again, the potential benefits of standardization should not be overlooked.

Supplier packaging normally travels only as far as to the manufacturing area, where its contents are unloaded and fed into the operations. At this point, three additional standardization factors come into play. One is that the mechanical means of movement, be it by fork truck, tow train, conveyor, or other means, should be standardized and the packaged loads standardized in accordance. Second, when materials are not moved directly to the manufacturing area, they must be stored. Storage conditions such as floor plan and storage racks should be standardized and the packaging should be standardized to be as compatible as possible. Third, the rate of production has an influence upon packaging standardization. Packaged supplies, ideally, should be of a content, size, and weight that minimizes the time during a production shift the line must be supplied. For example, if space in the manufacturing area permits the placement of eight hours worth of material, and that amount is arranged into a load that can be handled by the means of in-plant transport, only one move is required to supply an entire production shift's needs.

With in-plant handling of packaged materials, the maxim is "the less handling the better." That's because handling can cause product damage. It's been demonstrated that a fair amount of the damaged materials that reach the manufacturing process is not the direct result of faulty supplier workmanship, or even transportation-imposed damage, but rather the consequence of in-plant handling. Therefore, integrated standardization of packaging and in-plant handling conditions can reduce the chances of the manufacturing process receiving damaged materials.

There is a bright future for further automation of packaging unloading at the beginning of the manufacturing process and package loading at the end of the process. The application of robotics carries great promise, and a recent study on robotized packaging operations made by a leading university predicts that this type of handling will dominate in the coming decades. At the present time, however, robot handlers cannot match the flexibility and decision making of human handlers. Consequently, the packaging will have to help compensate by being standardized and held within tight tolerances.

The high-speed conveyorized movement of packaged goods through a facility is now claiming greater attention. That type of movement imparts vibrations and jolts that can cause product damage if the packaging has not been specified in recognition of that possibility.

Because so many manufactured products now contain electronic and computerized components, the influence of in-plant handling and packaging on the generation of electrostatic discharge is a prime area of concern. Those delicate, highly fragile components are easily rendered inoperable by charges produced when packages are opened or while products are being handled during assembly at the plant. Standardization of packaging materials, handling procedures, and controls such as grounding mats are called for.

DURABLE RETURNABLE CONTAINERS

A growing number of industries have gone to the use of durable, returnable containers in order to facilitate in-plant flows. Here, I'm referring to racks, bins, and totes—containers that are used repeatedly and that have a useful life measured in years. Under the right product and volume conditions, such packaging can provide substantial economies, not only in handling, storage, and inventory control, but also in transportation, assuming that the containers are not completely captive within the facility. Also, the disposal issue, so much a problem with expendable packaging, is all but eliminated. Because of the costs of these containers, standardization is difficult to overemphasize.

The basic trade-offs to be weighed involve initial capital investment, maintenance and repair, and administrative control of the containers.

SOME PRACTICAL CONSTRAINTS ON PACKAGE STANDARDIZATION

The first and foremost constraint is the systems approach within which the package standardization efforts should be performed. It's impossible to standardize in such a way that the positive effects on all concerned activities are maximized. Invariably, conflicts and the need for trade-offs will arise. For instance, the size, weight, and dimensions of packaged loads that are best suited for transportation economies may not yield the optimum economies in material handling.

A proven bit of advice: when tackling trade-offs, give priority to the least flexible elements of the system. Take, for example, transportation and storage. Transportation vehicles have inflexible dimensional and weight limitations, while storage racks are more manipulatable. It would probably be best to fit the packaged loads to the transportation vehicle requirements and then to seek out the most compatible adjustments in the more flexible areas.

Also keep in mind that too much standardization can hurt. Standardized packaging is not custom fitted to the product. Therefore, void fillers such as cushioning, padding, and dunnaging may be called for. The added cost can wipe out the benefits accrued from standardization. Relatedly, voids represent a reduction in packaged density. That translates into more handling and space requirements for a given volume of goods.

CONVINCING OTHERS

Perhaps, at this time, the reader is thinking about the potential difficulty of selling the standardization concept to others within the organization, particularly the marketing department. It's true that product proliferation and product diversity are long-entrenched marketing strategies. It would be extremely unwise to ignore competitive market requirements in specifying the characteristics of the product. Unwise, also, is the failure to attempt to meet market requirements, while simultaneously looking for manufacturing-related economies.

There are a number of important consequences stemming from too little standardization. Such mounting problems and trends as longer product (and packaging) development times, more product liability suits, and more frequent product recalls, are at least in part due to a lack of standardization. Clearly, the manufacturing firm should seek to know and quantify the economics of its present position on standardization, and to draft plans for improving these economies. There are various routes toward that pursuit. Among them are audits, supplier suggestions, customer suggestions, employee suggestion programs, and consultants.

The issue remains concerning how to go about developing a standardization program. By now, it should come as no surprise that I say that the approach should be systematic and integrated, as opposed to isolated and disjointed. I believe that packaging standardization should be made part of an overall program of product standardization.

That's easily justified since the product is the focal point of any packaging system. To the extent practical, the various products to be packaged should be developed or revised to eliminate all nonfunctional differentiating variables between them that affect the required characteristics of the packaging. When there are fewer pertinent packaging-related differences among the products, naturally there can be, as a consequence, fewer differences in the packaging used for the products.

Product standardization should be of great concern to manufacturing as well as packaging. The fewer the product variables that have to be managed and accounted for, the easier the manufacturing job. The relationship holds true even in this era of so-called flexible manufacturing systems and quick changeover production lines.

4

People

In an age fascinated by automation and computerization, the importance of people in the manufacturing realm can be undervalued. Actually, the human resource is more important than ever. Today and tomorrow, manufacturing and packaging professionals must meet the challenges of the modern-day industrial world. Building the foundation of education and experience required to support the nation's goals for reindustrialization—along with the automation and computerization this implies—will be a formidable task. Company survival means effective human resource planning and management: A company has to decide on the optimal number of people, their skill requirements, and how to keep employees trained and motivated.

THE ORGANIZATION

It makes sense to begin with a discussion on the influence of the organizational structure on people. The organizational structure is the arena in which people perform. Its characteristics directly affect performance.

Manufacturing's position on the organizational chart is fairly consistent from industry to industry. It is usually placed alongside the other core business activities, such as marketing and finance. Packaging, by contrast, has a less fixed position. While there are no universally agreed-to rules for packaging's position in a given company, its placement should reflect its perceived value as a function. For instance, packaging should not be positioned too low in the organization or subordinated to the wrong department.

There is a trend toward organizational structures developed around the interrelationships of key factors—costs, for example. The relatively new organizational concepts of Logistics, Material Management, and Physical Distribution follow that line of thinking. In a number of manufacturing firms, the packaging department is positioned under one of these areas. The most relevant concern is this: Does the department's placement in the organizational structure facilitate rather than encumber the way people work toward achieving the company's goals?

The following characteristics are typical of the most successfully organized departments:

1. A stated mission along with well-delineated objectives.
2. An emphasis on planning.
3. A structure that is compatible with the product and how it is sourced, manufactured, and distributed.
4. Specific policies and procedures governing internal and external interfaces.
5. The ability to acquire and retain quality people to carry out the activities of the department.
6. Adequate support in terms of budgeting and reserves.

THE PACKAGING PROFESSIONAL

The challenges of the manufacturing/packaging relationship require a special individual to meet them successfully. He or she must have good analytical, problem-solving, and communications talents. The individual must be, in effect, a combined visionary and realist, capable of creativity and innovation as well as hard-nosed, logical reasoning. Those qualities won't be found on every street corner.

Organizational problems could hamper the work of this rare professional. The perception that top management does not understand or appreciate the value of the packaging or manufacturing functions can be defeating.

I once originated a personal development program called TIP. The acronym stood for Talent, Image, and Performance. The program was based on the premise that a department's contribution is a function of:

- The quality of the talent it can attract to its ranks
- Its image
- Its performance

The three are interrelated. The greater the talent of the department's people, the better its image should—and must—be within the company. As a result, performance will be enhanced. A poor department image can deter talented people from joining it. A poor image can also sabotage performance if the poor image results in low expectations and tacit acceptance of mediocrity. Packaging and manufacturing can both benefit from a program designed to upgrade their talent, image, and performance.

THE ROLE OF POLICY

In managing the people side of the manufacturing/packaging relationship, as many particulars as possible should be formulated as policy. Developing a packaging department manual that details such things as the department's mission, organization, responsibilities, and so forth, along with guidelines defining the relationships between packaging and the departments it serves, is crucial. The manual should, for example, spell out packaging's responsibilities to manufacturing as well as that department's responsibilities to packaging. It should also delineate the communication and procedural channels through which the responsibilities are met.

It is essential that departments communicate during the development of the manual. By so doing, they can acquire understanding and respect for each others' responsibilities and concerns. This is important, since packaging serves a number of other departments and conflicts *do* arise. The potential conflict involving manufacturing/marketing requirements was referred to earlier. Since the packaging department sometimes has to act as a peacemaker, a manual that defines policies for resolving conflict can be of great help.

It is sometimes said that package development often overlooks manufacturing concerns. It may be true that some packaging personnel know a bare minimum about how the packaging they specify will be run at the plant. That can happen in companies where packaging is a centralized staff function, remote from the line operations located at distant plants. One solution is to have staff personnel serve a stint as line personnel, for six months to a year, and vice versa. Each would then see the world from the other side of the fence. The result—people who are better developed and versatile. Such cross-training can be made a policy requirement.

THE ROLES OF TRAINING AND MOTIVATION

Training has grown in importance because of the widening gap between managerial and technical requirements and the staff's inability to fulfill those requirements. There are two basic solutions: reduce the requirements or increase the ability. Training deals with the latter. Many companies discover that the poor skills of their personnel are the major obstacle to implementing state-of-the-art technology. Here, training becomes necessary. A properly trained workforce can increase productivity, lower production costs, reduce downtime, reduce overtime, improve product quality, produce less scrap and waste, and lower maintenance expenses, among other possible benefits.

All levels of packaging and manufacturing personnel, from management to production line operators, may require training. Regardless of who is being trained, the process should be approached as a cycle. The major steps are:

1. Verifying the need for training.
2. Establishing training objectives.
3. Determining informational requirements.
4. Developing training materials and a presentation format.
5. Deciding who will receive training.
6. Deciding who will conduct the training.

7. Performing the training sessions.
8. Evaluating and followup.

Consider, for example, how training can benefit the package line mechanic. Proper training enables the mechanic to determine, in the event of downtime, what went wrong and what to do about it—all with the minimum expenditure of time. The mechanic has become familiar with the line equipment as a result of his training, which included both classroom hours and hands-on experience with the equipment. Among other things, the mechanic has been trained to:

- Disassemble and reassemble the equipment.
- Trouble shoot.
- Build a malfunction into the equipment for diagnostic and correction purposes.

These and similar skills will be in great demand as manufacturing becomes more automated, and one of the major tasks for the remaining factory workers will be keeping the automated systems maintained and working.

Managers should also be trained. Managers in the manufacturing and packaging areas need to sharpen their skills in long-range planning, financial analysis, and worker psychology, to cite a few examples.

Motivation is closely associated with training—the ability to perform is often a function of the desire to perform. Low motivation among workers can be costly, resulting in poor productivity and work of inferior quality. Less obvious indicators of low motivation are buried in worker turnover rate, absenteeism and—in extreme situations—dissatisfaction to the point of sabotage. Even psychologists disagree as to how a company should motivate its workers. On certain points, however, they do agree. They are:

1. Motivation efforts should be targeted at encouraging a specific behavior in the worker.
2. Motivation is multidimensional. A worker can be motivated by many different drives, each of which has more or less importance at different stages in his or her life.
3. Incentive systems must have direct relevance to the worker in order to be important to him or her.

4. Motivation attempts have their greatest chance of succeeding when the worker is evaluated by standards that are unambiguous, fair, and quantifiable; when the worker is kept abreast of his or her performance through clear and timely feedback; and when performance-related rewards and penalties are applied equally to all.

EDUCATION

Although there has been progress on the academic front, much still remains to be done. Currently, packaging curricula across the country are emphasizing manufacturing subjects, and business curricula are adding manufacturing courses. Well-educated personnel with an understanding of manufacturing and packaging will be a boon to industry. Not that all manufacturing and packaging personnel have to be college-trained. Both manufacturing and packaging, however, have often suffered as a result of poorly trained personnel who lack the educational credentials of their counterparts in other departments.

Universities lag years behind industry in important areas such as computer-driven technology. High costs prohibit the purchase of adequate equipment. Thus, in their courses, universities resort to simulations or small-scale versions of the systems. Industry can help form a closer college/industry coalition by donating equipment and sponsoring university projects.

5

Getting Equipped

Will the United States achieve its reindustrialization goals? Only if our manufacturing operations are performed faster, more efficiently, and more economically. Much of our success will be determined by the equipment and machines we use.

Top management must take the leadership role in technology investment. After all, investments in facility improvements *are* long-term capital expenditures. Given the prevailing negative attitude toward the long term, the following factors must be faced squarely:

- The emphasis on short-term returns and the quarterly and annual profit-and-loss statements has produced myopia and fear.
- Few managers are willing or able to take the calculated and intelligent risks associated with investments. Most feel that their advancement is tied to the results they can generate in the near term.
- As a result, the success of many companies has been mortgaged at the price of attractive annual income reports.

The responsibility rests with top management to reverse these attitudes by:

- Recognizing the importance of long-term investments in technology and facilities improvement.

- Rewarding managers who demonstrate foresight—those who can generate investment proposals which reflect a bird's-eye view encompassing the company's long-term survival.

With encouragement from the top level, middle managers of functional areas like manufacturing and packaging can be imaginative and creative in exploring opportunities to equip their facilities for the future.

THE AUTOMATED FACTORY: A PLACE FOR PACKAGING

"Reindustrialization" is now viewed as industry's remedy for its manufacturing woes. The "automated factory" will be the setting in which it will all take place. Exactly what is an automated factory? It is best described as a *concept* or goal, and not as a place or an achieved fact—because automation is acquired in degrees and levels, and is therefore relative. My definition of the "automated factory" is as follows: a manufacturing/production environment whose characteristics and composition are determined by pursuit of the long-range goal of achieving the optimal (not necessarily the maximum) degree of automation. In other words, the concept is an integrative one. It ties together the various functions of manufacturing into one operating system and allows for the optimal level of automation required to run and control the system.

Contrary to popular belief, the automated factory is not a world without people, operated solely by computers and machines. No facility is fully automated. The idea involves, rather, a carefully planned and engineered move toward greater automation. In automating a factory, management faces formidable decisions as to which equipment, systems, and controls to combine. Automation experts agree that while there is no model for the automated factory, all have these common characteristics.

1. Production planning and scheduling systems.
2. Manufacturing machines and operations.
3. Material handling equipment.
4. Information processing and transmitting systems.
5. Control systems.

The automated factory emphasizes the need to integrate manufacturing and packaging efficiently. Because the automated factory depends on the interface of various automated systems, the packaging provided must be compatible with manufacturing operations. That means packaging must be more standardized; in many instances, stronger and of better quality; and subject to tighter controls and tolerances. This is particularly true of the material handling system, which is responsible for moving goods and components through the plant. Many of the systems that deliver materials to the manufacturing floor, transfer materials from manufacturing, and take materials into storage depend on packaging for their best performance.

HOW PACKAGING WILL HELP RUN THE AUTOMATED FACTORY

Packaging and containerization that support automated manufacturing have to be compatible with the equipment. Compatibility is important in manual methods as well. The following examples will serve as illustrations.

Automated Storage and Retrieval (AS/R). In the first stage of an AS/R feasibility study, the unit loads that the system will store and retrieve are classified. The more similar the loads, the easier it is to design an AS/R for them. The dimensions and weights of each load must conform to the dimensions and weight-bearing capabilities of the storage slots of the AS/R. This match-up determines storage density, a true measure of AS/R efficiency. Loads that bulge too much or otherwise exceed established dimensional limits can cause problems. Electronic sensors "measure" incoming loads to make sure they are not too large to fit into the storage slots. Hence, how well the load is assembled and held together within allowed tolerances is critical to good AS/R performance.

Automatic Guided Vehicle System (AGVS). An AGVS is a driverless, horizontal transporter of unit loads. The loads carried have to be secure, stable, and within stated dimensional limits so that they will not tilt over, come apart, or otherwise prove inconvenient or unsafe. These qualifications are largely a function of how the loads are unitized (palletized, film-wrapped, strapped, contained in racks or bins, and so on)—all packaging considerations.

Robotics is perhaps the most widely publicized technology in the automated factory. Earlier, I mentioned that robots will increasingly perform more of the package loading and unloading within the manufacturing environment. One such operation involves the robotic loading of bottled lotion into partitioned corrugated shipping cartons. Replacement of the carton requires constant control. Because the robot lacks the vision and adaptability of a person, the corrugaged carton has to be placed in practically the same position each time and the dimensions of the partitioned cells must be held within tolerances so that the preprogrammed placement of the bottles into the carton is achieved without mishap.

Computer tracking and accountability of materials from receival through shipping is another feature common to the automated factory. Bar code technology makes it possible. A number of packaging factors affect a bar code system: the quality of the code printing, the contrast between the code and its background color, and the shape of the package, to name a few.

The automated factory will also employ more conventional types of handling methods such as industrial trucks, conveyors, and cranes. However, packaging improvements will be important here as well. In order to better match the weight of unit loads, adaptations of the fork truck, such as push-pull attachments for slip sheets and clamp attachments for palletless loads, should reflect the types of packaged loads that the equipment must handle. In all events, the compatibility between the loads and the handling mechanisms influence the cost, efficiency, and safety of the move. From a stricter manufacturing point of view, compatibility means potentially less handling-induced damage to the materials transported into and away from the manufacturing operations.

Packaging and unit loads are most efficiently moved along a conveyor when their sizes and weights match the width and power-drive mechanism of the conveyor. When cranes are used, there is an added element of danger while the load is temporarily suspended in mid-air. If the load becomes unattached from the slings or the bottom of a crate falls out, serious accidents can occur.

AUTOMATING THE PACKAGING LINE LAYOUT

An efficient *packaging line* represents the intelligent choice of equipment to best accommodate the manufacturing scheme. This holds true regardless of the degree to which the facility is automated.

Packaging lines should be planned and laid out to provide a reliable daily account of line performance. The planning should anticipate problems related to worker assignments, downtime and maintenance, and output measurements.

Manufacturing faces the dual pressures of cost control and improved productivity. The packaging line layout can influence the overall productivity strategies by reducing the quantity of labor required (through automation), by raising the skills of the remaining workforce, by increasing production speeds, and by controlling the costs of packaging materials.

The physical appearance of a packaging line can take a number of forms. Whatever its configuration, every assembly line should:

1. Be compatible with the nature of the product.
2. Have the output capacity to accommodate current conditions.
3. Have the capacity to expand output for forecasted future conditions.
4. Be planned in consultation with the operating and maintenance personnel, since they will have the greatest contact with the line.
5. Ease the supplying of material at the front of the line and the removal of material from the end of the line.
6. Be economical to service and maintain.
7. Have the correct degree of versatility and flexibility.
8. Have equipment and machinery components that are smoothly integrated.
9. Be economical in labor requirements.

10. Be efficient in energy usage.
11. Comply with environmental considerations.
12. Comply with safety requirements.
13. Meet budgetary and financial justification hurdles.
14. Facilitate the necessary quality control and inspection procedures.
15. Facilitate the collection of data required to evaluate overall line performance.
16. Be a balanced reflection of the company's marketing, manufacturing, engineering, financial, and packaging concerns.

THE BASICS OF PACKAGING MACHINERY

Choosing machinery is a complicated decision. One must consider the conditions of numerous vendors, the wide choice in types of machines, changing technology, and optional features. The decision, again, calls for a systematic approach: (1) define your situation, (2) formulate requirements on the basis of your situation, and (3) choose machinery on the basis of those requirements.

Defining the situation

The saying "You can't know too much about your business" applies when assessing and defining your situation. If you want to make the most intelligent and informed decisions for your company, you should know thoroughly your company's characteristics—its products, suppliers, markets, competition, and projections for growth.

Formulating your requirements

The second step requires precision. Too often, machinery requirements are stated in vague, general terms. This shortcoming need not occur if the manager follows the suggestions mentioned in the last paragraph. Two common consequences of nonspecific requirements are overinvestment and underinvestment. You have overinvested when the machine capabilities exceed your needs for the present and the foreseeable future. You have underinvested when the machine capabilities fall short of your needs. Given the long-term nature of capital investment, both situations are costly in their occurrence and in their correction. The specific nature and priorities of requirements will differ from company to company. Nonetheless, the following are category headings on which to base machinery requirements.

1. *Investment.* The price ceiling set by the budget; financial justification hurdles (for example, payback and discounted cash flow); means of acquisition (buying, rental, or leasing); the acceptable terms of payment; the importance of investment credits and other tax incentives.
2. *Labor.* The number of workers required; their levels of skill; the interchangeability of job classifications; the interchangeability of male and female workers; training requirements.
3. *Speed.* Top speed vs. optimal speed.
4. *Flexibility.* Changeover; the number of different packages or materials to be run; fixed-position vs. mobile equipment.
5. *Maintenance.* Frequency; ease; spare-parts inventory requirements.
6. *Technology.* State-of-the-art or something less; custom made or "off the shelf."
7. *Safety.* Union and regulatory requirements; requirements that simply make good sense; environmental requirements such as emissions and noise-level controls.
8. *Performance.* Production up-time; quality of output; consistency; track record.
9. *Service.* Guarantees and warranties; quickness of dispatch.

Choosing the machinery

This step is made easier by the two previous steps. Easier, but not easy. Some alternatives will have been eliminated because they will not meet the formulated requirements, but the remaining choices may still be wide. It is difficult—perhaps impossible—to know the entire range of packaging machinery markets. The individual company should follow a programmed collection of information on specific categories of equipment. The technical press, seminars, technical associations, trade shows, and vendor communications are cost-effective ways of becoming and remaining knowledgeable about packaging machinery.

There are several trends in the packaging machinery markets that the decision-maker should be aware of. One concerns *affordability*. Machine manufacturers are attempting to bring their products within the affordable reach of wider markets. Manufacturers are not exactly engaging in price wars, and in most instances, you get what you pay for. Still, the prices of some machine categories are becoming reasonable.

Another trend is toward *simplification:* of design, of operation, of maintenance. Yet another is the move to *systems integration* through advanced computerized electronics. Machines can stand alone or be linked by electronics to one or more other machines in a packaging line.

The last trend involves the changing makeup of the machine supply markets. Many suppliers are diversifying. For some, that means selling the materials that run on the machines they supply. For others, it means diversifying into other kinds of machines that are typically teamed in packaging lines. Whether this "supermarket" direction is good or bad is a matter of viewpoint. A company buying new machinery must form its own opinions and devise its action plans accordingly.

COMPUTERIZED CONTROLS

Experts inform us that despite the tremendous strides already made, the possibilities of computer applications in manufacturing are far from being exhausted. Primary examples include:

The Programmable Controller (PC)

A notable computer application in the packaging and manufacturing equipment area is the programmable controller. PCs allow a degree of production-line integration not possible in the past. The individual components of the line can be controlled and linked to other line components by the advanced electronics of the PC. The result is control through on-line real-time information generated by the sensing, monitoring, and correcting capabilities of the computer.

The payoffs are enormous:

- Production problems are revealed as they occur, rather than afterward.

- Corrections can be made more quickly and at lower cost.
- Product or material changes can be made easier and faster with push-button panels.
- Diagnostics and trouble shooting can be performed more effectively and efficiently.

PCs give managers access to reports of a strategic nature. For instance, systems can be programmed to generate information on line speeds, when and where stoppages occur, and the percentage of rejects. The reports can be used to determine:

- The location of the most critical components on the line.
- Which components are contributing most to bottlenecks and downtime.
- Which revisions are worth considering to create a packaged product at the desired cost, quality, and level of efficiency.

The Computer Hierarchy

Most categories of packaging machinery are available with programmable capabilities, and many previously acquired machines can be retrofitted and integrated into a computerized operation. In considering computerized controls, managers must decide on the amount of linkage between individual machines as well as the structure of the facility's computer hierarchy. Hierarchy refers to the levels of control that make up a control network. It works like this: A mainframe host computer resides at the highest level and controls production and scheduling. It is linked to a second-level computer responsible for operations control and the development of daily schedules. The second level is linked to a third-level minicomputer that controls the production machines and equipment. The individual machines receive their commands through their PCs.

STANDARDS AS CONTROLS

Standards are measuring devices. They are a means of control that permit managers to determine if the object being measured is below, at, or above acceptable levels of quality established by the standard. Once this has been determined, management can take the appropriate action. A standard can be developed for nearly anything. The central concern should be that it is fair, relevant, and capable of being correctly applied. The importance of standards for packaging was discussed earlier in this briefing. However, there are other standards that impact on the relationship of packaging and manufacturing.

Labor standards. These are customarily devised by industrial engineers using such tools as the time study. Any point of contact between a worker and the packaging can be evaluated with the intent of devising a standard for it. How incoming packaging materials are unloaded, handled, stored, and fed into the production system can be made the objects of standards. Standards make it possible to apply consistent methods that yield better control and protection of the packaging materials. The materials produced can, in turn, perform better under manufacturing conditions.

Standards for machinery. Standards, when applied to machinery, can provide:

- Control over changeover operations.
- Control and monitoring of production rates.
- Surveillance to facilitate maintenance and repair.
- More accurate production forecasts.
- More precise machine loading.
- More accurate work assignments.

Increasingly, machine standards are being based on information gathered through computerized monitoring and control systems, as discussed in the preceding section.

Standards for procedures. Using the procurement function as an example, standards should be applied to the following:

- The procurement of packaging materials.
- The procurement of machinery.
- The procurement of supplies.
- The assignment of responsibility for evaluating procurement needs.
- The vendor selection process.
- The development and issuing of specifications.
- Follow-up audits.

Gearing up for packaging today's—and tomorrow's—products by using the tremendous potential made available by the electronic revolution calls for careful consideration of all the factors discussed in this chapter. From here, we move to the phase that caps the manufacturing/packaging process—the terminus of our continuum—getting the packaged product into the hands of the industrial/retail consumer.

6

After the Manufacturing Is Performed

There are no distinct lines of demarcation that permit one function to be fenced off from those functions that are systematically related to it. Nor can one conclude that concerns for manufacturing and its integrated packaging operations stop at the end of the production line. The product has been manufactured and packaged for one purpose: to get it in the possession of the customer, whether industrial or consumer. An entire set of activities, commonly known as physical distribution, form the third link of the sourcing-manufacturing-distribution continuum that composes all industrial based systems.

The integrity with which the product has been manufactured and packaged comes to light under the rigors of the distribution system. There are often decisions to be weighed regarding how much strength and protection should be manufactured into the product itself, versus how much product protection should be afforded by the packaging. Regardless of the trade-offs made, the objective is for the customer to receive the product in a satisfactory condtion. Continued sales and profits are the rewards to the company that continually makes that happen.

PREPARATION FOR SHIPPING

There are certain elements of packaging operations strictly devoted to preparing the product for shipment. Casing and unitizing equipment are examples. Other activities that can be mechanized and integrated into the manufacturing process include case farming, palletizing, slip sheeting, shrink and stretch wrapping, strapping, bundling, taping, gluing, and labeling.

Not all such packaging operations are mechanized, however. Nor should they be. Shipping crate set-up, filling, and closing are mostly a manual endeavor, as is the loading of many types of durable containers: wire-mesh containers, rackers, and steel corrugated bins, for example. Even there, the operations can be studied for areas to improve labor efficiency.

The nature of the distribution packaging methods used affects the products' susceptibility, not only to physical damage and deterioration, but also to loss through misidentification or by pilferage. The performance of distribution packaging affects manufacturing. If the packaged product is lost or stolen during distribution or arrives damaged beyond acceptance, the effective cost of manufacturing those products that do arrive satisfactorily increases.

For some industrial products, it may pay to investigate the possibilities for bulk packaging. Triple-wall corrugated boxes, industrial bulk bags, and metal bulk bins have the potential of lowering the packaging costs per unit, reducing handling costs, and lowering transportation charges. The growth of bulk packaging has been limited in recent times because of the investment in filling equipment and handling equipment they often require.

In making decisions about distribution packaging and the equipment it entails, there are some basic considerations to honor. In the case of the industrial product manufacturer, many of the considerations are those discussed in Chapter 3 under the subtitle, "Standardizing Supplier Packaging." The only difference is that the manufacturer is now the supplier and stands subject to the packaging and shipping requirements of his customer.

The consumer product manufacturer may face a more complicated set of conditions in that the customers are more numerous and the distribution channels more varied. Studies have conclusively shown that many packaging practices do not reflect the requirements of physical distribution. Case counts may be out of line with typical order quantities, for example, or unit loads may result in poor utilization of transport vehicle space, or the packaging may have poor handling features. Without going too far afield from strict manufacturing considerations, I do want to make it clear that part of the manufacturing and packaging relationship is devoted to preparing the product for shipment to the customer. When the customer receives a satisfactory product, his payment reimburses the incurred manufacturing and other production-related costs. And customer satisfaction smoothes the way for the repeat of profitable transactions.

7

A Parting Look

Economists, marketers, and philosophers have, in their individual ways, stated that mankind's wants are insatiable. The proof of that claim continues to be reflected in the desire for products. Products that provide comfort, convenience, prestige, safety and other real or imagined benefits are demanded in ever-increasing numbers by increasingly prosperous societies worldwide. Manufacturers compete to cash in on this virtual bonanza. The most successful ones are, and will be, those who can manufacture the products under optimal conditions of cost, productivity, and quality.

This briefing has been devoted to the manufacturing function and its interface with the packaging function, particularly the way the two interact to contribute to a company's success. Packaging is indispensable to manufacturing—it affects the quality of the components that go into the manufacturing line operations. And it affects the delivered quality of the product by protecting it throughout distribution to the customer. The interface, therefore, operates along the entire industrial system.

As the world moves into a space age of automation and computerization, the challenges placed on manufacturing will be demanding but, at the same time, exciting. Every function and activity related to manufacturing will have to be better integrated into a systematic unit. Manufacturing people will need to be more knowledgeable about packaging; and packaging people will have to be more knowledgeable about manufacturing. I would hope that this briefing has made a small contribution toward that end by highlighting some of the major interdependencies between the two functions.